D1218713

The Rosen Publishing Group's

New York

Published in 2006 by The Rosen Publishing Group, Inc.
29 East 21st Street, New York, NY 10010



Book Design: Jennifer L. DiPasquale

Photo Credits: Cover, p. 1 © M. Thonig/H. Armstrong Roberts; p. 4 © Dick Dickinson/International Stock; p. 7 © Steve Hix/FPG International; p. 11 © Bill Terry/Viesti Associates, Inc.; p.12 © Ken Reid/FPG International; p. 14 © Corbis.

ISBN: 1-4042-3343-1

Library of Congress Cataloging-in-Publication Data

Richter, Al.
Ocean tides / Al Richter.
p. cm.– (The Reading room collection. Science)
Includes index.
ISBN 1-4042-3343-1 (lib. bdg.)
1. Tides–Juvenile literature. I. Title. II. Series. III. Rosen Publishing Group's reading room collection. Science.
GC302.R53 2006
551.46'4--dc22

2005011889

Manufactured in the United States of America

# Contents


Ocean Facts 5

What Are Tides? 6

The Moon and Tides 9

Deep Tides 10

Animals and Tides 13

People and Tides 14

Glossary 15

Index 16

# Ocean Facts

Oceans are very large bodies of salt water that cover much of Earth. Oceans supply most of the water that makes clouds and rain.

Many animals, like fish, clams, and **lobsters**, live in oceans. Many plants live in oceans, too.

Water covers almost three-quarters of Earth.

# What Are Tides?

**Tides** are the **constant** rise and fall of the oceans every day. Tides make waves that come in and go out slowly during the day.

High tide is when the waves reach high up on the beach. Low tide is when the waves reach only a little way up the beach.

Tides happen in oceans all over the world.

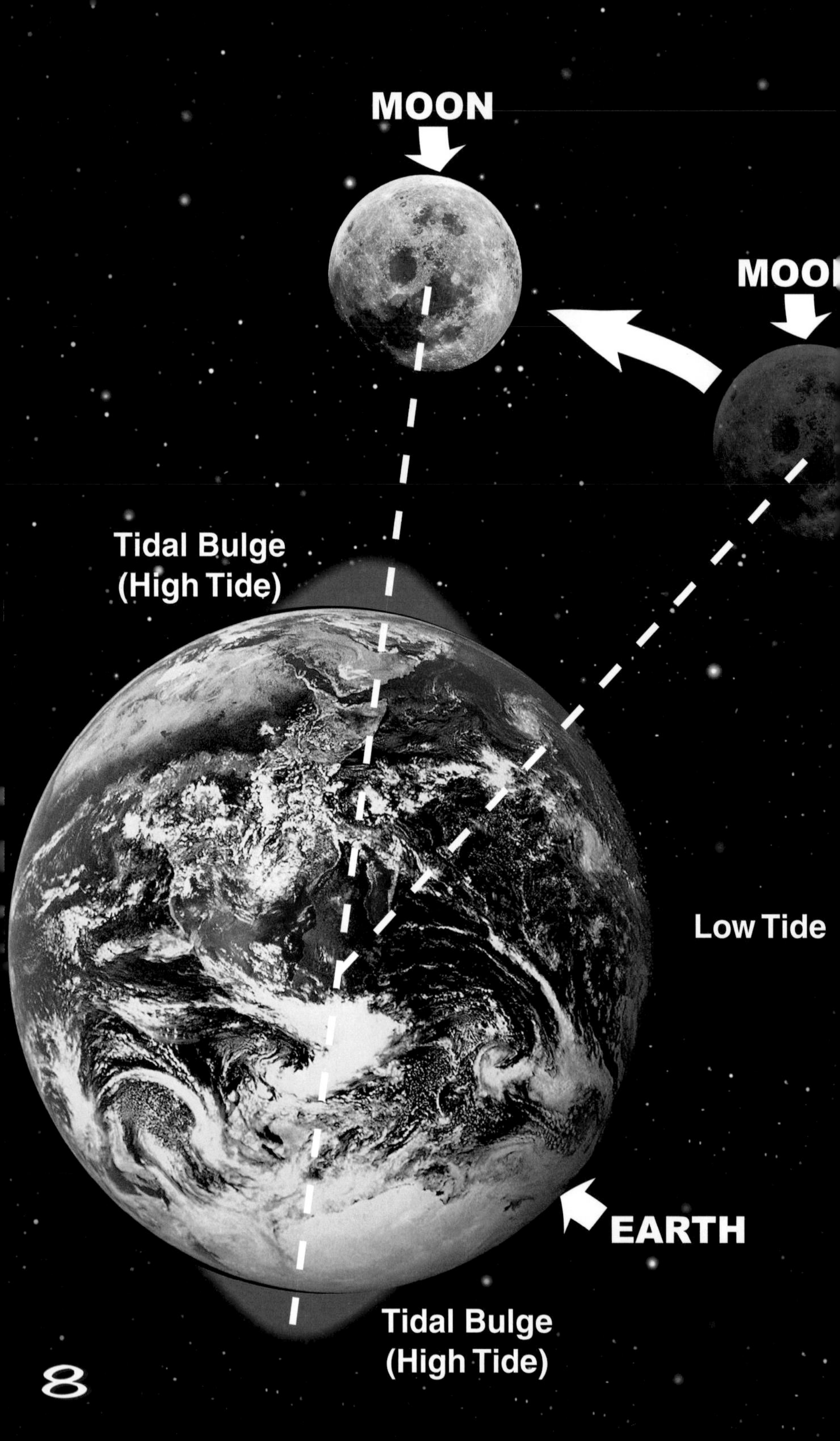
MOON
MOO
Tidal Bulge
(High Tide)
Low Tide
EARTH
Tidal Bulge
(High Tide)

# The Moon and Tides

The moon **circles** Earth and helps to make tides by pulling on the water closest to it. This makes a **tidal bulge**, or rise, in the water. The water on the other side of Earth also forms a bulge. These bulges are what cause high tides. When the bulges move on, low tides happen.

The tidal bulges in oceans move slowly around Earth as the moon does.

# Deep Tides

Tides usually cover a beach when they come in and uncover it when they go out. When tides rush into a narrow area, the water cannot spread out. The difference between high tide and low tide can be great in these places.

At high tide, the water reaches all the way to the tops of these rocks!

# Animals and Tides

Animals that have special parts live in water where there are tides. Some animals have thick, hard shells to keep them safe from crashing waves. Some stick to rocks and docks to keep from being carried out deeper into the ocean. Some animals hide deep in the sand to stay safe.

Starfish have tiny feet with suckers on them. The suckers help them stick to rocks when tides come in and go out.

# People and Tides

People who fish for a living watch the tides so they know the best times to fish. Towns near an ocean can **flood** if the tide comes in when there is a **storm**.

We watch the ocean tides for many reasons. We are learning more about tides every day.

# Glossary

**circle** To move around something else.

**constant** To never stop.

**flood** To cover or fill with water.

**lobster** A shellfish with ten legs, with large claws on the front pair.

**storm** A strong wind with rain, snow, or hail.

**tidal bulge** A rise in the ocean caused by the pull of the moon.

**tide** The daily rise and fall of oceans.

# Index


## A

animals, 5, 13

## C

circles, 9
constant, 6

## F

flood, 14

## H

high tide(s), 6, 9, 10

## L

lobsters, 5
low tide(s), 6, 9, 10

## M

moon, 9

## O

ocean(s), 5, 6, 13, 14

## P

people, 14

## S

storm, 14

## T

tidal bulge, 9